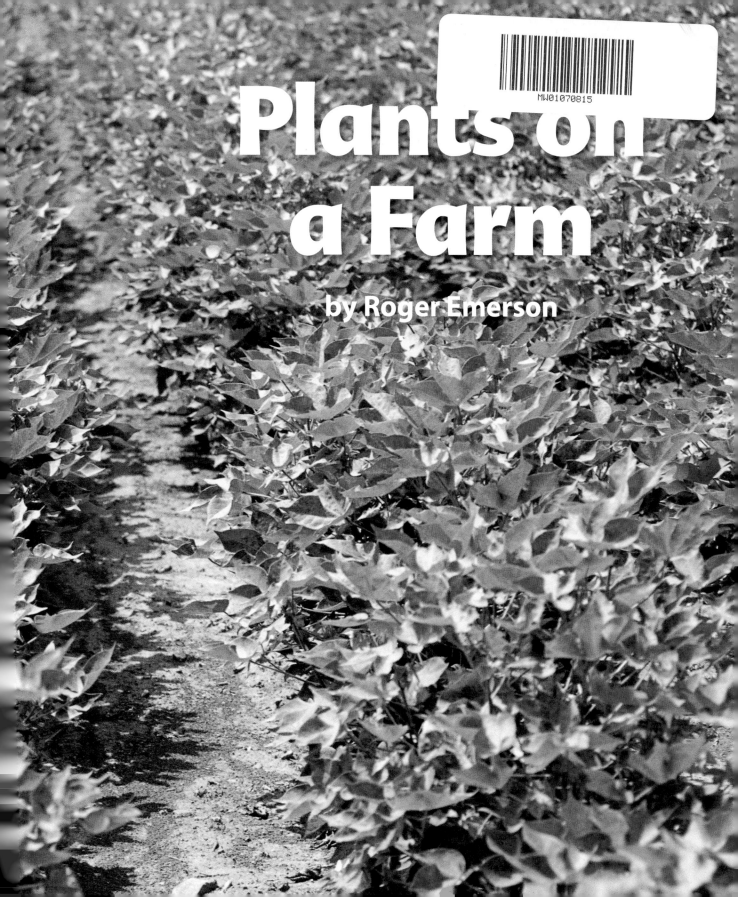

Plants on a Farm

by Roger Emerson

Lettuce plants have **leaves.**

Cabbage plants have leaves, too.

Corn plants have **stems.**

Wheat plants have stems, too.

Watermelon plants have seeds.

seeds

Tomato plants have seeds, too.

seeds

Carrot plants have **roots.**

root

Radish plants have roots, too.

root

Strawberry plants have **flowers.**

Pumpkin plants have flowers, too.

Plants on a Farm

leaves

roots

stem

flower

seeds